# Manual of Radiotelephony for Airline and other Pilots

Capt. Philip A. Ikomi, Ph.D.

Copyright © 2008 by Philip A. Ikomi



ISBN  0-7414-4882-3

*Published by:*

**INFINITY**
PUBLISHING.COM

*1094 New DeHaven Street, Suite 100*
*West Conshohocken, PA 19428-2713*
*Info@buybooksontheweb.com*
*www.buybooksontheweb.com*
*Toll-free (877) BUY BOOK*
*Local Phone (610) 941-9999*
*Fax (610) 941-9959*

*Printed in the United States of America*

*Printed on Recycled Paper*

*Published May 2009*

# Table of Contents

## <u>Preface</u>

This book is intended both as a ready reference for seasoned pilots, and a guide to new airline pilots who need to perform at the level expected in the highly demanding air traffic situation at busy and rapidly growing airports around the world. I was motivated to write this book as a result of my experience in the airline both as a Training Captain and an Instrument Rating Examiner. A lot of the pilots joining the airline in which I operated lacked formal radiotelephony (R/T) training and were consequently poor material for the bulk of their co-pilot duties, which was mainly communication. It then dawned on me that I could write a book in which I would spell out the basic communication requirements; and as much as possible, present the real world picture of R/T. This is the culmination of that idea.

The book is quite straightforward and easy to use. The reader is first introduced to the tools of the trade. Then I have gone on to describe the contents of a typical R/T message. Following this general description, I then quickly went on to a simulation of a typical journey from starting engines to landing at a destination airport. In the mean time, not so normal flight occurrences are simulated, while all through the book simulated messages are explained where it is felt that an explanation might enhance the learning process. Throughout the book, there is an emphasis on making the transmissions as real as possible since it is felt that the usefulness of a technical manual lies in how well the materials learned from the book can be applied to practical situations. In communication, much of what one perceives depends on what one expects to hear. Without this expectation, one may not understand what is said, no matter how clearly the words sound.  A visit to an auction sale or the race tract will attest to the verity of my last statement.  There, although the words are in English, you may not understand a word because you have no expectation of what to hear.  What I have done in this book is emphasize the things that you should expect to hear so that when you enter the airspace, you will find that you are able to understand much of the conversations from the very first minute.

I hope that you will find this book quite helpful in all phases of flight and that you would remember a pilot maxim, which says, "Pilots help the next fellow" and so recommend it to your colleagues.

# Alphabets

For the clarity of expression required in transmissions on radiotelephone (R/T), the alphabets are given one-word designations.  Letters that are not clear to the recipient when pronounced are spelled using the following words that make up the international radiotelephone alphabets used by all private and commercial aircraft pilots:  The phonetic pronunciation is in (parenthesis)[1].

A  -  Alpha (AL FAH)
B  -  Bravo (BRAH VOH)
C  -  Charlie  (CHAR LEE OR SHAR LEE)
D  -  Delta (DELL TAH)
E  -  Echo (ECK OH)
F  -  Foxtrot (FOKS TROT)
G  -  Golf (GOLF)
H  -  Hotel (HOH TEL)
I  -  India (IN DEE AH)
J  -  Juliet  (JEW LEE ETT)
K – Kilo  (KEY LOH)
M  -  Mike (MIKE)
N  -  November (NO VEM BER)
O  -  Oscar (OSS CAH)
P  -  Papa (PAH PAH)
Q  -  Quebec  (KEH BECK)
R  -  Romeo (ROW ME OH)
S  -  Sierra  (SEE AIR RAH)
T  -  Tango (TANG GO)
U  -  Uniform  (OO NEE FORM)
V  -  Victor  (VICTAH)
W  -  Whiskey (WISS KEY)
X  -  X-ray (ECKS RAY)
Y  -  Yankee (YANG KEY)
Z  -  Zulu (ZOO LOO)

## **Numerals**

1 – One (WUN)
2 – Two (TOO)
3 – Three (TREE)
4 – Four (FOW-er)
5 – Five (FIFE)
6 – Six (SIX)
7 – SEV-en (SEV-en)
8 – Eight (AIT)
9 – Nine (NIN-er)
0 – Zero (ZEE-RO)
Decimal point – Decimal (DAY SEE MAL)

Note that when referring to numbers with decimal points, you are to pronounce only the word "Dayseemal". Do not include the word "point". In addition, note that numbers are pronounced as individual component numerals. The FAA wants you to say "point" but would honor the ICAO use of "Dayseemal" in place of point[2].

Examples:
14 - One-four  (WUN FOW-er)
150 - One-five-zero  (WUN-FIFE-ZEE-RO)
1600 - One-six-zero zero (WUN SIX ZEE RO-ZEE RO  or
        (WUN-SIX-HUNDRED)
25,000  -  two-five-thousand (TOO-FIFE-TOUSAND)
27879  -  Two-Seven-Eight-Seven-Nine (TOO-SEV-
        en-AIT-SEV-en-NIN-er)
121.5 – (WUN-TOO-WUN DAYSEEMAL FIFE.)

When numbers are in whole hundreds, thousands, or millions, they are pronounced with the word "hundred", "tousand", or "million", respectively, after the individual numerals have been enunciated.

Further Examples:

100  -  One-hundred (WUN HUNDRED)
1500 - One-five-hundred  (WUN-FIFE-HUNDRED)
43,000 -  four-three-thousand  (FOW-er-TREE TOUSAND)
38,000.000  -  three-eight-million (TREE-AIT-MILLION)

In the US National Airspace System (NAS), airway or jet route numbers are pronounced as follows:[3]

V10 – VIKTAH TEN
J435 – J FOW-er THIRTY-FIFE

## Readability Scale[4]

The strength of the transmitted message in terms of clarity, is subjectively scaled.  The scale is subjective in that each recipient of a message decides whether or not the message is clear.  Signal strength is rated from one to five, five being the most effective while the lower end of the scale terminates in one.

| | |
|---|---|
| 5 | - Very effective |
| 4 | - Not so effective |
| 3 | - Intermittent |
| 2 | - Hardly readable with a lot of background noise |
| 1 | - Not readable or unreadable |

If you read somebody 3, for example, it is advisable to ask for a repetition of each word.  You would make this request in the following manner:

"I read you 3 (TREE), words twice, over."

On the other hand, if the signal strength is only 1 or 2, then ask for a complete repetition by saying:

4

"You are unreadable, say again, over."

In another instance you may have read part of the transmission and not read other parts.  For instance, you may have flown through some static that partially blanked off the transmission.  In that case, you do not wish to hear the whole message again, so you should say:

"Say again all after…, over"

or

 "Say again all before…, over,"

depending on what part of the message you want repeated.  If you wanted to hear the message again, you would say:

"Say again your last transmission, over"

or

"Say all again, over."

## Radio Check

Time was when standard procedures were used to make a radio check.  In those days, you would say:

"Preflight check on WUN, WUN, AIT DAYSEEMAL WUN (118.1 MHZ).  How do you read, over?"

These days you should make a normal communication with the Air traffic Control facility (ATC) to get some pertinent information.  This will serve the additional purpose of checking your signal strength and that of the station you are calling.  Procedurally[5], you should do the following:

Select the desired frequency, listen for any on-going conversations, and when satisfied that you will not interrupt a conversation in progress prior to your tuning your radio to that station, call, giving the following information in the order shown below:

1.	The call-sign of the ATC facility on the selected frequency, e.g., Nino tower.
2.	Your call-sign, e.g., NO VEM BER WUN SIX TREE HOH TEL (N163H).
3.	The nature of your message or request, e.g., "Request time check." In the standard procedure you would request "preflight", "signal", or "readability" check on the selected frequency at this stage, stating the frequency explicitly, enunciating the numbers "1, 2, 3, 4, 5" and asking how you are read.
4.	The word "over"

An example will make this point clearer. Assume your call-sign is NO VEM BER WUN SIX TREE HOH TEL and that you are talking with Nino tower. Your message will be transmitted according to the numbered sequence below:

1.	Nino tower
2.	NO VEM BER WUN SIX TREE HOH TEL
3.	Request time check
4.	Over

For the standard procedure, it will be necessary to substitute the following for item 3 above.

3.	Readability check on WUN, WUN, AIT DAYSEEMAL WUN (118.1 MHZ) WUN, TOO, TREE, FOW-er, FIFE. How do you read, over?

This standard procedure is not strictly adhered to however, simply because it is time consuming in our crowded air

space today where verbal communication is being drastically reduced to the barest minimum.
In response to your request for time check, Nino tower may say the following :

"Roger, NO VEM BER WUN SIX TREE HOH TEL, ATC time ZEE RO AIT FOW-ER FIFE (08:45)."

To the readability check, the tower may respond in this manner:

"Roger, NO VEM BER WUN SIX TREE HOH TEL, I read you FIFE, FOW-er, TREE, TOO, or ONE, out"--depending on applicable signal strength.

## Time Reporting

Coordinated Universal Time (UTC) is used world wide in aviation. The letters UTC stand for the French equivalent of Coordinated Universal Time hence the odd looking representation UTC rather than CUT. It was called Greenwich Mean Time (GMT) based on the time on the meridian over Greenwich in England. It is still essentially the same time. It is also known as Zulu time, and denoted by the letter Z. When reporting it is not necessary to say you are using Zulu time or UTC time because unless otherwise stated, it is assumed to be Z-time.

Since we want to be as brief as possible in order not to congest the frequency, we usually report only the minutes past each hour when the party being addressed understands the hour we are referring to. This is usually the current hour. For instance, if when over a radio facility, "KD" you reported 08:35, and your next reporting point "KA" was only twelve minutes away, you would report it as "FOW-er SEV-en" (47) representing the minutes past eight and not the entire 08:47. However, when you are giving your estimates, each time your estimate changes to a new hour, you must state the whole four digits of the time. For example, consider a journey from point A to point B, which has five reporting points for which estimates are required. These points and their estimates for a hypothetical journey are as follows:

| Depart | A | 08:10 |
|---|---|---|
| Estimate (est) | TJ | 08:55 |
| Est | KK | 09:20 |
| Est | RK | 09:35 |
| Est | IN | 10:07 |
| Est | BN | 10:28 |
| Arrival | B | 10:49 |

In reporting the above estimates, you need only say the following:

Departed <u>Alpha,</u> ZEE RO AIT WUN ZEE RO, estimate <u>TANG GO JEW LEE ETT</u>, FIFE, FIFE, <u>KEY LOH, KEY LOH,</u> ZEE RO NIN-er TOO ZEE RO, <u>ROH ME OH KEY LOH</u>, TREE FIFE, <u>IN DEE AH NO VEM BER,</u> WUN ZEE RO ZEE RO SEV-en, <u>BRAH VOH NO VEM BER</u> TOO AIT, arrival <u>BRAH VOH</u>, FOW-er NIN-er.

Note that with a few exceptions, in virtually all transmissions involving letters, the word equivalent of the alphabets are used. Hence "Alpha" , "Tango", … "November" used above in place of the corresponding alphabets.
To more realistically bring home the use of radiotelephone to you, we shall now describe the basic requirements of a typical message and then simulate a typical journey during which we shall introduce all conceivable messages that may occur in real life.

## <u>Contents of a Typical R/T Message</u>

A typical message consists of four (4) elements.
a.      The call-sign of the station being addressed.
b.      The call-sign of the calling station.
c.      The substantive message being transmitted
d.      The word "over" or "out", which ever is applicable.

Example:

a.      Nino tower (the station being addressed)
b.      NO VEM BER WUN SIX TREE HOH TEL (the calling station)
c.      Request taxi clearance (the substantive message)
d.      Over.

## Normal Position Reporting[6]

In a normal position report, item "c" in the typical R/T message above, would include four parts that have to be given in a particular sequence.

These are: (1) Position on ground over which you are. This could be a Very High Frequency Omni-directional Radio Range (VOR) facility, an intersection of VOR radials, abeam a non-directional beacon (NDB), or over an NDB. The position could even be given by the geographical coordinates on an Inertial Navigation System, (INS), Global Positioning System (GPS), Loran C, etc. (2) Your time at the position, (3) Your altitude, and (4)Your next position and estimated time at that position and optionally, (5) the next position after that. Specific ordering of the information is required because it aids reception when the ATC controller knows what to expect to hear. It could be quite confusing otherwise. In addition to these, you will include what is called PIREPS or Pilot Reports of Meteorological observations. These are ordinarily, winds aloft, temperature, and turbulence factor as experienced by you, but may also include icing where present in significant amount, thunderstorm, hail stones, etc.

Example:

(a)      Nino tower
(b)      NO VEM BER WUN SIX TREE HOH TEL (N163H)
(c)      1. KEY LOH BRAH VOH (KB), 2. ZEE RO AIT TREE FIFE (08:35), 3. Flight level TREE, TREE ZEE RO (FL330), 4. Estimating KEY LOH AL FAH (KA) FOW-er SEV-en (0847)
(d)      Over.

10

With additional information, ( c ) above will become:

( c ) (1)  KEY LOH BRAH VOH  (2) ZEE RO AIT TREE FIFE (3)  Flight level TREE. TREE. ZEE RO (4) Estimating KEY LOH AL FAH FOW-er SEV-en.  (5) HOH TEL NO VEM BER next  (6)  Wind, ZEE RO NIN-er ZEE RO, diagonal SEV-en FIFE (090/75)  (7)  Temperature, minus TREE FIFE degrees (-35 degrees Celsius)  (8)  Moderate chop (Turbulence).

Radiotelephony is like telegraphic messages in that it is very precise.  Although I have used letters, Arabic and Roman numerals in outlining the steps in message transmission, it should be noted that these messages are not discrete when read, but transmitted in a continuous string.  For example,

"Nino tower, NO VEM BER WUN SIX TREE HOH TEL, request taxi clearance, over."

## <u>Establishing Communication Contact</u>

When you put on a radio station frequency for the first time, you are to listen to any on-going conversation before knowing where to start your own conversation with the station. You should also not key your mike until you know there is room for you to be heard. In addition, you should not key your mike because if there is an on-going conversation you will not be heard even if you keyed your mike. Instead, you may be blocking a transmission that is on-going. Therefore, always listen first once you have tuned your radio to the frequency. Then, when calling a station for the first time, you should call the station and wait for its response before passing your message.  This ensures two things – ( 1) You are heard and  (2)  You hear the station. Thus, after your selection of the appropriate frequency, you should have the following initial dialogue, assuming you were NO VEM BER WUN SIX TREE HOH TEL (N163H):

**N163H:**   Nino tower, NO VEM BER WUN SIX TREE HOH TEL (over)

**Nino Tower:**  NO VEM BER WUN SIX TREE HOH TEL, Nino tower, go ahead (over)

With this introductory contact you can then go on to the main body of your communication.  Note that in conversations with any ATC facility, if you just read back whatever is said, you cannot go wrong.  That is also a very good way to improve on your R/T since the controllers are invariably trained in R/T.  The congestion that may result will be tolerated as other pilots in the vicinity will realize that you are a novice, and need to be given a chance to operate freely.  Pilots help the next fellow.  Flying is sometimes a matter of life and death.  Standardization of communication in flying has a military basis with precision and clarity as the watch words.  Hence there is not much room for inconsistency and ambiguity.  It is thus advisable that you use the same words and phrases that will appear in this book in your own transmission.  It is the same phraseology that is used throughout the world.  English is also the international aviation language.  We are now in a position to simulate the various stages of flight communication.

## Start-Up

**N163H**:  Nino tower, NO VEM BER WUN SIX TREE HOH TEL, request airfield data (over)

Note that there is no Automatic Terminal Information Service (ATIS) for departures (ATIS Departures) at this airport otherwise, this request would not have been made.  Note also that the request for start up would be directed to

Ground Control in larger airports with such facilities. It is such airports that usually have ATIS also.

(Note also that in the above request, the word "over" is put in parenthesis to indicate that it need not be said since it is understood. This will be the practice for the rest of the book.)

**Nino Tower:** Roger, NO VEM BER WUN SIX TREE HOH TEL, altimeter WUN ZEE RO WUN WUN  millibars (1011 mb.) (or Hectopascals), temperature, TREE ZEE RO degrees Celsius (30°C), surface wind, ZEE RO FIFE ZEE RO degrees, diagonal WUN FOW-er knots (050/14 kts), time check, ZEE RO AIT ZEE RO TOO (08:02), runway ZEE RO FIFE (R/W 05) in use.

**N163H:** Roger, Nino tower, altimeter WUN ZEE RO WUN WUN (1011) hectopascals, temperature, TREE ZEE RO, runway ZEE RO FIFE, time checked, NO VEM BER WUN SIX TREE HOH TEL.

It is pertinent that you always read back all numbers transmitted whether or not read back of the numbers is asked for. For example, notice that in the above exercise, the altimeter setting (QNH), temperature, and runway were read back. However, time is not read back but an indication is given the controller that the time had been checked since the time would ordinarily have changed even as the process of read back is underway.

Note however that ATC will give you time to the nearest quarter minute to enable you set your time. Fractions of a quarter minute less than 8 seconds are given as the preceding quarter minute while fractions 8 or more seconds are given as the succeeding quarter minute.[7]

Notice the use of two digits – zero (0) and five (5) for the runway designation above. It is standard International

Civil Aviation Organization (ICAO) practice to use two digits for runway designations. However, in the United States, there is a tendency to use only one number where the first number is a zero. For instance, runway ZEE RO FIFE (R/W 05) would be designated Runway FIFE (R/W 5).

Observe that it is customary to give your call-sign whenever you transmit. You may give it either at the end as was done above, or at the beginning of the transmission following the station you are addressing, as is done in the start up request below.

**N163H:** Nino tower, NO VEM BER WUN SIX TREE HOH TEL, request start up clearance.

**Nino Tower:** Roger, NO VEM BER WUN SIX TREE HOH TEL cleared to start, temperature TREE ZEE RO degrees Celsius.

**N163H:** Cleared to start, temperature TREE ZEE RO degrees Celsius, NO VEM BER WUN SIX TREE HOH TEL.

Note that pertinent information for starting engines is ambient temperature. Hence the tower gives it when clearance for start up is requested even though the aircraft did not ask for it.

Having started the engine(s) it is time to taxi out for take-off; hence a request for taxi clearance will be made.

**N163H:** Nino tower, NO VEM BER WUN SIX TREE HOH TEL, request taxi clearance.

**Nino Tower:** NO VEM BER WUN SIX TREE HOH TEL, Nino Tower, you are cleared to taxi via the outer to runway ZEE RO FIFE; altimeter, WUN ZEE RO WUN WUN, surface wind, ZEE RO FIFE ZEE RO, diagonal WUN SIX knots.

Whenever there is a significant weather, additional information may be passed along with the taxi clearance. For instance, if there is icing or snow, the thickness may be given; if there has been recent rain, the tower may inform you that the runway surface is slippery and advise you to taxi with caution.  If surface visibility is below a certain minimum, Runway Visual Range (RVR) rather than the surface visibility (depending on the country) is given and here too a cautionary advice is given.

**N163H:**  Roger NO VEM BER WUN SIX TREE HOH TEL, cleared taxi, runway ZEE RO FIFE via the Outer, altimeter, WUN ZEE RO WUN WUN  Hectopascals.

In some African countries, the number of souls on board the airplane as well as the fuel endurance is required by ATC. This information is passed to them during the taxi or just as you make the request for taxi clearance.  In addition, during the taxi phase of your flight, ATC in some African countries would pass you ATC clearance.  You are therefore required to be ready to copy such a clearance.  In Europe and the United States of America, clearance is obtained from a Clearance Delivery frequency and is usually passed while aircraft is still on chokes and on your requesting permission to start engine(s).

**Nino Tower:**  NO VEM BER WUN SIX TREE HOH TEL, Nino Tower, ATC clears  NO VEM BER WUN SIX TREE HOH TEL as filed, JEW LEE ETT TREE (J3) departure, flight level SIX ZEE RO (FL 60) or above at ROW ME OH intersection, squawk AL FAH TREE ZEE RO TREE TREE (A3033) with CHAR LEE after takeoff, NO VEM BER WUN SIX TREE HOH TEL.

**N163H:**  Roger, ATC clears NO VEM BER WUN SIX TREE HOH TEL as filed, JEW LEE ETT TREE (J3) departure, flight level SIX ZEE RO (FL 60) or above at ROW ME OH intersection, squawk AL FAH TREE ZEE RO

TREE TREE (A3033) with CHAR LEE after takeoff, NO VEM BER WUN SIX TREE HOH TEL.

**Nino Tower:** NO VEM BER WUN SIX TREE HOH TEL, Nino tower, that is correct.

If ATC has reason to suspect that your R/T is not sound she may specifically request that you read back her clearance, by simply adding the words "read back": at the end of the transmission, thus:

**Nino Tower:** NO VEM BER WUN SIX TREE HOH TEL, Nino Tower, ATC clears NO VEM BER WUN SIX TREE HOH TEL as filed, JEW LEE ETT TREE (J3) departure, flight level SIX ZEE RO (FL 60) or above at ROW ME OH Intersection, squawk AL FAH TREE ZEE RO TREE TREE with Charlie after takeoff, read back, over.

Since you have been taxiing and doing your taxi checks and copying clearance, you should be about reaching the runway. You will therefore have to request take-off clearance so that you do not lose time holding at the holding point especially if there is no traffic ahead of you.

**N163H:** Nino Tower, NO VEM BER WUN SIX TREE HOH TEL, request take-off clearance.

**Nino Tower:** NO VEM BER WUN SIX TREE HOH TEL, Nino Tower, you are cleared for take-off runway ZEE RO FIFE. After take-off, contact Nino Approach on WUN TOO TREE dayseemal WUN (123.1).

In places where Standard Instrument Departures (SIDs) are not used, take-off clearance would be more elaborate, e.g., Lagos, Nigeria as illustrated by the following example.

**Lagos Tower:** NO VEM BER WUN SIX TREE HOH TEL, Lagos Tower, you are cleared for take-off, runway WUN NIN-er. After take-off, maintain runway heading until passing TREE FIFE ZEE RO, ZEE RO feet (3,500 ft), thereafter turn left, heading WUN FOW-er ZEE RO degrees (140 degrees) till out of flight level WUN, WUN ZEE RO (FL 110); thereafter resume normal navigation, over.

The response will then be:

**N163H:** Roger cleared for take-off, runway WUN  NINer; after take-off, maintain runway heading till passing TREE FIFE ZEE RO, ZEE RO feet, thereafter turn left heading WUN FOW-er ZEE RO degrees till out of flight level WUN, WUN ZEE RO, thereafter resume normal navigation, NO VEM BER WUN SIX TREE HOH TEL.

The aircraft now takes off and radio frequency switched to Nino Approach  (123.1 MHz) as required in the take-off clearance from Nino Tower.  In Western Europe and the United States, you may not switch to the assigned frequency until tower clears you to switch.  In most parts of Africa and the Middle East, ATC will contact you on Tower frequency, give you your airborne time, and the Approach Radar or Approach Control frequency and ask you to contact either one.

## Western Europe/USA

We have assumed that there is no ATIS departure at this airport of departure.  However, in Western Europe and the United States, most large airports have this service.  Hence it is imperative to deal with what happens in those places before we continue with the simulated flight.  We shall assume the take-off to be at a large and busy airport too.  These requirements attach further responsibility on the pilot.  In such larger airports there is a Ground Control frequency to handle all ground traffic.  Ground is usually responsible for issuing parking instructions after landing, clearance delivery, start up, pushback, and taxi clearances.  Tower issues take off clearances.  When requesting start up clearance in a large airport like London Heathrow, you need to tell ground control that you have received their ATIS departure.  This knowledge is needed to ensure that you have the latest terminal information being broadcast.  The ATIS usually has call –signs for the various times of broadcast, each new broadcast being an update of an earlier one.  ATIS usually contains weather information as well as information pertinent to air navigation at that particular airport.  For instance, a typical ATIS broadcast may be like this:

"This is Heathrow information alpha for ZEE RO AIT ZEE RO ZEE RO zulu.  Surface wind, ZEE RO FIFE ZEE RO diagonal …southern portion of the taxi way to runway TOO AIT closed to traffic, aircraft to …"

When you call Heathrow ground for start up clearance, you are therefore expected to say that you have received information alpha.  Because it is possible that if you were ready for start up an hour or two after receiving "information alpha" the information may have altered considerably at the half-hourly updates, you need to check the current ATIS just prior to making your start up request.  If you did not check, and there has been only a minor change since you got "Information Alpha", when you make your start up request, ground will update your information.  If on the other hand there has been a major change, when you request start up ground may refer you back to get the current ATIS departure, especially if she is very busy and has no time to amend your information. (Note that many large airports operate a Tower Data Link System (TDLS) that includes a Pre-Departure Clearance (PDC) function that automates clearance delivery operations to participating users. Aircraft with the facility receive this information via the Aircraft Communications Addressing and Reporting System (ACARS) or similar data link system. Aircraft that do not have ACARS could still receive this information through a printer at the departure gate.[8])

In addition to stating that you have received the departure information by stating exactly which one you have received, you will, at large airports, have to specify the stand where you are parked and your destination. Let us look at an example using London Heathrow Airport.

**N163H:** Heathrow ground NO VEM BER WUN SIX TREE HOH TEL, **kilo two**, with "alpha", destination Kano, request start up clearance.
Note that "Kilo two" is the stand where you are parked and "alpha" refers to the eight o'clock departure ATIS at Heathrow.

With large airplanes, or even with small transports at airports where they have been parked on stands that use telescopic avionic gangways or next to terminal buildings, it is customary to request permission to "pushback". Pushback refers to the process of pushing an aircraft backward from the building to position it for taxi after engines are started. Most aircraft do not move backwards on their own power under normal circumstances so they have to be pushed backward. It is also normal to start one engine and then request pushback and while being pushed backwards, the other engines are started. This ensures that all engines are not started close to the building with all the attendant noise and vibrations. Remember that good airmanship demands that you are one step ahead, so while starting the first engine, make sure that you will be ready to request pushback immediately after that engine gets started. The pushback request is a simple one, but you also have to give ground control your location; so include it in your request for pushback thus:

**N163H:** Heathrow ground, NO VEM BER WUN SIX TREE HOH TEL, kilo two, request pushback.

When you get your ATC clearance, which in Europe or the United States you get while still on chokes, as mentioned earlier, ATC could address you as "Heavy" in addition to your call-sign if yours is a big transport such as the B777, B747, DC-10, etc as indicated in your flight plan. In England, B747 pilots are advised to use the word "Jumbo" in all their communications. This assures you that ATC is aware that you may not be able to comply with her speed limitations during flight at low altitudes where a maximum of 250 knots is mandatory. The A380 will similarly be treated as it is much bigger than the B747.

Let us now continue in our trip with what obtains in Western Europe during the departure phase. Just airborne, your SID directed you to switch frequency to **Nino Approach** and you just switched. Since you are under radar surveillance, you just listen and maintain your visual look out for traffic after switching frequencies.

**Nino Approach:** NO VEM BER WUN SIX TREE HOH TEL, radar contact.

At this stage ATC could make amendments to the clearance as traffic demands, or she could give you vectors, that is, headings to steer. Here is an amended clearance:

**Nino Approach:** NO VEM BER WUN SIX TREE HOH TEL, radar contact, on reaching flight level six ZEE RO, maintain; report at Romeo Intersection for further climb, over.

**N163H:** Roger, maintain flight level six ZEE RO, report at Romeo Intersection for further climb, NO VEM BER WUN SIX TREE HOH TEL.

## Africa/Middle East

Recall that it was said that in Africa you are likely to be contacted on the tower frequency by the ATC after you are airborne and be given your airborne time. Notice how this is done in the following transmission:

**Nino Tower:** NO VEM BER WUN SIX TREE HOH TEL, Nino tower, airborne WUN ZEE RO (08:10) contact approach radar now on WUN TOO TREE dayseemal WUN (123.1 MHZ) over.

Notice that you have to read back the time in this case and observe that the frequency is read back too since it is made up of numbers.

**N163H:** Roger, WUN ZEE RO, approach radar, WUN TOO TREE dayseemal WUN, NO VEM BER WUN SIX TREE HOH TEL.

The tower may or may not acknowledge the foregoing reply. If she is busy, she probably is already giving clearance to some other traffic. So do not wait for her to reply, but go on to the assigned frequency. If your read back was wrong, she is likely to respond correcting you. At busy airports, with a new pilot unfamiliar with ATC procedures, the tendency is to get confused when ATC does not acknowledge the read back but gives another frequency to some other traffic departing toward another sector a minute or less after you. You now think that the new message was intended for you and you scramble for the new frequency at the same time as you try to read back the putative correct version of your clearance. By this time, the legitimate owner of the clearance is also trying to read back. The result is that nobody is heard as the frequency is jammed leading to delays. Maintain a listening watch at all times to save you from such an embarrassing situation. Remember also that

ATC always calls the call-sign of an aircraft along with any clearance or, shall we say, communication with each aircraft. Thus if you listen closely, you are not likely to mistake another's clearance for your own.

In an African country with radar service, the journey we started continues like this:

**N163H:** Nino Radar, NO VEM BER WUN SIX TREE HOH TEL, WUN TOO TREE dayseemal WUN (over).
Note that the frequency 123.1 MHZ was given earlier along with your airborne time after take-off.
**Nino Radar:** NO VEM BER WUN SIX TREE HOH TEL, Nino Radar, radar contact, continue as cleared.

Radar control may make pertinent adjustments to the original clearance. In fact, at every stage of your flight, ATC has the right to amend your clearance to ensure safety; so be ready to comply.

Since in Africa SIDs are not common, we shall continue the simulated flight using the elaborate clearance mentioned earlier—the clearance from Lagos Tower. Additionally, in Europe and the United States, communication on R/T is usually kept at the barest minimum during departure because of radar usage (and of course, it is safer to do so) hence, we need only emphasize what obtains in places with less reliance on radar. We shall however tell you what to expect in Europe and the United States. Even in Africa there is a difference in procedures between places where radar service is available and where it is not. With radar service, not much talking is necessary since your activities are being monitored. However, it is prudent to report to keep ATC reminded of your traffic. Additionally, after you are airborne you may report thus:

**N163H:** Nino Approach, NO VEM BER WUN SIX TREE HOH TEL, climbing out of TOO FIFE ZEE RO, ZEE RO feet (2500 ft.) on runway heading.

22

**Nino Approach:**  Roger, NO VEM BER WUN SIX TREE HOH TEL, Nino Approach.

There is nothing special about the altitude 2500 ft.  It just happens to be where you were at the time you made the transmission.  Announcing this passing altitude to ATC makes them aware of your position as you make your initial transmission on the new frequency.  You should also note that the above conversation is strictly not necessary as you have been sighted on ATC radar.  However, the following conversation is necessary:

**N163H:**  Nino approach, NO VEM BER WUN SIX TREE HOH TEL out of TREE FIFE ZEE RO, ZEE RO feet, turning left to WUN FOW-er ZEE RO degrees, over.

**Nino Approach:**  Roger, NO VEM BER WUN SIX TREE HOH TEL, Nino approach, call leaving flight level WUN, WUN ZEE RO and setting course.

The necessity for the foregoing conversation lies in the fact that it affords the ATC another opportunity to correct you if your maneuver is wrong.  In this case, a left turn is correct.

**N163H:**  Nino Approach, NO VEM BER WUN SIX TREE HOH TEL, leaving flight level WUN, WUN ZEE RO, turning on course (over).

**Nino Approach:**  Roger, NO VEM BER WUN SIX TREE HOH TEL, Nino approach, standing by for your estimates.

Observe how the estimates are given below and compare them with what you learned earlier.

**N163H:**  Nino approach , NO VEM BER WUN SIX TREE HOH TEL, estimating Akan, FOW-er FIFE (:45), Kano FIR (Flight Information Region) (boundary) FIFE AIT (:58),

Papa Hotel (PH), ZEE RO NIN-er WUN WUN (09:11), destination Iko, TOO ZEE RO (09:20) (over).

**Nino Approach:** Roger, NO VEM BER WUN SIX TREE HOH TEL, estimates copied, call again level at flight level TOO NIN-er ZEE RO (FL290).

In the mean time, you will try to contact Kano on High Frequency (HF) since you are required to contact Kano at least ten minutes prior to entering the Kano airspace. Nino Approach is a hypothetical location in the Republic of Benin. This ten-minute requirement is universal for most FIRs. On North Atlantic crossings, twenty minutes are required prior to entry. Most FIRs coincide with International boundaries except where some countries, because of their small sizes fall into a larger country's FIR.

Transmission on HF is slightly different from that on Very High Frequency (VHF) because of the noise from static interference attendant on HF communications. For this reason, it is expedient to repeat the call-sign of the station you are calling a minimum of twice when establishing communication contact. Each time you wish to transmit a message, you first have to call the facility, get their response, and then talk. In VHF, you are expected to maintain a listening watch whereas in HF, you are not because of excessive noise level associated with HF. However, you are expected to give your *selcal* (selective calling[9]) call-sign as soon as you have established initial contact. This facility will enable the ground station call you in case there is a message for you. Thus it is expedient that ground station should give you a functional check of the *selcal* equipment by activating it in a check soon after your initial contact with the ground station. You therefore have to request this check from the ground station on initial contact. In HF communication it is also necessary to state the frequency on which you are calling the ground station since the ground operator has only one speaker for a number of frequencies. For instance, Kano has only one speaker system for a number of HF frequencies—8826 KHZ, 5519 KHZ, etc. In addition, the ground station will give you its primary and secondary frequencies. You should guard the primary frequency and if you lose contact on the primary, be sure to first try the secondary. In transmissions on HF it is necessary to call only the first two digits of the HF frequency as an abbreviation of the whole number. Having said that, let us now proceed with the simulated flight and establish contact with Kano.

**N163H:** Kano, Kano, NO VEM BER WUN SIX TREE HOH TEL, calling Kano on AITY AIT (8826) (over).

**Kano Center:** NO VEM BER WUN SIX TREE HOH TEL, NO VEM BER WUN SIX TREE HOH TEL, Kano Center, go ahead.

**N163H:** Kano, Kano, NO VEM BER WUN SIX TREE HOH TEL, from Nino to Iko, departed Nino, ZEE RO AIT WUN ZEE RO (08:10), flight level TOO NIN-er ZEE RO (FL290), estimating Akan, FOW-er FIFE (08:45), Kano FIR, FIFE AIT (08:58), PAH PAH HOH TEL, ZEE RO NIN-er WUN, WUN (09:11), destination Iko, TOO WUN (09:21). SIX SEV-en souls on board, endurance remaining ZEE RO TREE WUN ZEE RO (03:10). Selcal, BRAH VOH KEY LOH WISS KEY FOKS TROT, request selcal check (over).

**Kano Center:** Roger, NO VEM BER WUN SIX TREE HOH TEL, estimates copied. Call again Kano FIR maintaining flight level TOO NIN-er ZEE RO, check selcal (over). (The ground station now activates your *selcal*.)

The *selcal* check consists of an alert in the form of a chime, audible notes, and a red light depending on the on-board equipment. A button in the flight deck is pressed to silence the audio, chime, and light, and a response is then made to the check.

**N163H:** Roger, Kano, Kano, NO VEM BER WUN SIX TREE HOH TEL, selcal check okay, will call you FIR, maintaining flight level TOO NIN-er ZEE RO.

We now return to Nino Approach to continue with VHF reporting.

**N163H:** Nino Approach, NO VEM BER WUN SIX TREE HOH TEL, Akan, FOW-er FIFE (09:45), flight level TOO NIN-er ZEE RO, estimating Kano FIR, FIFE AIT (09:58), PAH PAH HOH TEL next.

**Nino Approach:** Roger, NO VEM BER WUN SIX TREE HOH TEL, Nino Approach, call again Kano FIR.

**N163H:** Roger, NO VEM BER WUN SIX TREE HOH TEL.

**N163H:** Nino Approach, NO VEM BER WUN SIX TREE HOH TEL, Kano FIR, FIFE AIT (09:58), flight level TOO NIN-er ZEE RO (FL 290), estimating PAH PAH Hotel, WUN ZEE RO WUN, WUN (10:11) (over).

**Nino Approach:** Roger, NO VEM BER WUN SIX TREE HOH TEL, call HF contact Kano (over).

**N163H:** Nino Approach, NO VEM BER WUN SIX TREE HOH TEL, HF contact Kano (over).

**Nino Approach:** Roger, NO VEM BER WUN SIX TREE HOH TEL, continue with Kano on HF, good day (out).

**N163H:** Kano, Kano, NO VEM BER WUN SIX TREE HOH TEL, calling Kano on AITY-AIT (8826 Khz) (over).

**Kano Center:** NO VEM BER WUN SIX TREE HOH TEL, Kano, go ahead.

**N163H:** Kano, Kano, NO VEM BER WUN SIX TREE HOH TEL, FIR, FIFE AIT, flight level TOO NIN-er ZEE RO, estimating Papa Hotel, WUN ZEE RO WUN, WUN, Iko next. Winds, ZEE RO TOO ZEE RO diagonal TREE FIFE, temperature, minus TOO FIFE (-25°C) (over).

**Kano Center:** Roger, NO VEM BER WUN SIX TREE HOH TEL, Kano, estimates copied, wind, ZEE RO TOO ZEE RO diagonal TREE FIFE (W/V 020/35), temperature minus TOO FIFE (-25°C) Call VHF contact Iko (over).

You now select Iko Approach frequency and make a call. Note however, that what you have learned here in respect of HF transmission to Kano is largely true of Santa Maria, New York, or Shanwick control HF communication. Certain regulations do however apply to oceanic crossings; some are discussed later in this book.

As you switch to your terminal approach frequency, it is again expedient that you indicate your type of aircraft when you make your first transmission to ATC so that she can separate you from other aircraft especially if you are a heavy transport. If you are a large transport for instance, you may be a turbulent wake hazard for lighter aircraft. On flights to the United States, large aircraft like the DC-10's and the B747's or B777's use the word "heavy" as they do in departures. On London flights, the word "Jumbo" is preferred. This is also very important because as you will recall, the speed restrictions in force during take-off mentioned earlier, apply to the approach when airplanes arrive back at low altitudes for the landings. Therefore, this is your chance to let them know that you are a big bird and may not be able to maintain some slow speeds on the approach. Remember however, that professionalism prevents you from saying that you are not able to comply with speed limitations when you are light and thus able. Check this example out:

"Kennedy control, NO VEM BER WUN SIX TREE HOH TEL heavy, Sates at TREE ZEE RO, flight level TREE FIFE ZEE RO with BRAH VOH, top of descent."
Note the use of "top of descent" here rather than the more traditional "request descent". They mean the same thing. Note also that this communication is brief unlike the one we shall see when we continue with our simulated flight because of the fact that Kennedy is in the United States, and there is radar coverage. Observe too that you stated that you have information Bravo, the arrival ATIS, which you obtained prior to requesting descent. In order to further aid your memory, we shall assume that your airplane is a large transport as we continue below.

**N163H:** Iko Approach, NO VEM BER WUN SIX TREE HOH TEL heavy, WUN TOO FOW-er dayseemal TREE (124.3) (over)

Note that by calling the station, Iko Approach and saying the frequency on which you are calling, you are matching the station to the frequency so that if that is not the case, you will be corrected immediately by the station whose frequency you called. For instance if that frequency does not belong to Iko Approach but Iko Radar, Radar will immediately refer you to Iko Approach.

**Iko Approach:** NO VEM BER WUN SIX TREE HOH TEL heavy, Iko Approach, go ahead.

**N163H:** Iko Approach, NO VEM BER WUN SIX TREE HOH TEL heavy from Nino to Iko, departed Nino, ZEE RO AIT WUN ZEE RO (08:10), flight level TOO NIN-er ZEE RO (FL 290), check Kano FIR, FIFE AIT (08:58), estimating PAH PAH HOH TEL, ZEE RO NIN-er WUN, WUN (09:11), destination Iko, TOO ZEE RO (09:20). SIX SEV-en souls on board, remaining endurance, ZEE RO TOO TREE ZEE RO (02:30).

Note that the departure time is read in full—08:10—since it was definitely already past the hour of eight at the time of this communication; while at Kano FIR you only report the time after the hour already mentioned. Again the full time 09:11 is given for the next point where the hour has changed and the destinations given in minutes past the time already mentioned since there is now no ambiguity. It is not really necessary to say "remaining endurance" as any endurance reported at this stage has to be what is actually left. All you need say is "endurance".

Over great distances such as across deserts and the high seas where navigational facilities are huge distances apart, and in parts of continental Africa where there are no airways, a tag is placed on the souls on board and endurance in case of search and rescue. HF is also used on such routes because VHF range depending on line-of-sight transmission, is too short. With the abundance of airways in Western Europe and continental United States of America, VHF is used and endurance and souls on board are not asked for on R/T. (In the United States, VHF frequencies are boosted to have extended range so that within the continental United States, HF frequencies do not have to be used. Such extended range VHF channels are said to be *remoted* VHF channels.) As most countries in Europe and the United States have adequate radar coverage, position reports are not often demanded.

**Iko Appraoch:** NO VEM BER WUN SIX TREE HOH TEL, cleared IN DEE AH KEY LOH VOR flight level TOO NIN-ner ZEE RO (FL 290), no delay, runway TOO FOW-er (R/W 24), KEH BECK NO VEM BER HOH TEL (Altimeter setting) WUN ZEE RO WUN TOO Hectopascals (1012) (over).

(Hectopascals is a measure of barometric pressure now in use in much of the world).
**N163H:** Cleared IN DEE AH KEY LOH VOR, flight level TOO NIN-er ZEE RO (FL 290), no delay, runway TOO

FOW-er (R/W 24), altimeter, WUN ZEE RO WUN TOO Hectopascals (1012) (over).

**Iko Approach:**  NO VEM BER WUN SIX TREE HOH TEL, Iko Approach, that is correct.  Standby for the nine o'clock weather.

Where Arrival ATIS frequency exists, the weather will be broadcast on it.  It is usually a VHF frequency and it is known as VOLMET broadcast when it is carried by a VOR facility.  You will need to know these frequencies in advance of your preflight .  For a long distance flight, there are HF ATIS frequencies.  Make a note of them in your navigation Manuals for easy reference.  Let's continue with the trip communications by acknowledging to standby for the weather.

**N163H:**  Roger, NO VEM BER WUN SIX TREE HOH TEL.

In the mean time, we can call Kano on HF to report contact with Iko on VHF.
**N163H:**  Kano, Kano, NO VEM BER WUN SIX TREE HOH TEL on AITY-AIT (8826 Khz).

**Kano Center:**  NO VEM BER WUN SIX TREE HOH TEL, Kano, go ahead.

**N163H:**  Kano, Kano, NO VEM BER WUN SIX TREE HOH TEL, VHF contact Iko  (over).

**Kano Center:**  NO VEM BER WUN SIX TREE HOH TEL, Kano, continue with Iko on VHF.

**Iko Approach:**  NO VEM BER WUN SIX TREE HOH TEL, copy the Iko NIN-er hundred weather (09:00):  Surface wind, TOO ZEE RO, ZEE RO, diagonal WUN FIFE (W/V 200/15), visibility, FIFE kilometers in mist, FOW-er octas

(4/8) at SIX, SIX, ZEE RO (660) meters, TREE octas (3/8) at TREE TOUSAND (3,000) meters, SEV-en octas (7/8) at NIN-er TOUSAND (9,000) meters, KEH BECK NO VEM BER HOH TEL (QNH) WUN ZEE RO, WUN TOO (1012) HectoPascals, temperature, TOO TOO degrees Celsius $(22^{o}C)$ (over).

**N163H:** NO VEM BER WUN SIX TREE HOH TEL, weather copied, KEH BECK NO VEM BER HOH TEL, WUN ZEE RO WUN, TOO (1012), temperature TOO TOO; call you ready for descent.

**Iko Approach:** Roger.

## Oceanic Control Area[10]

Twenty minutes prior to entering a designated Oceanic Control Area (OCA) you are required to contact the appropriate ground ATC facility for your Oceanic clearance. Most likely, you will be communicating with them on HF. The following rules apply to Oceanic crossing points.

1.      All times should be transmitted in four digits, e.g., ZEE RO AIT WUN ZEE RO (08:10) rather than WUN ZEE RO only in normal VHF transmission.
2.      The word "position" must be included in a position report.
3.      In making position reports, report your present position, your next position, and the position after that.
4.      Your Mach number must be included in your position report if a Mach number was part of your clearance.
5.      On Westbound flights, you are to squawk transponder A3000 and on Eastbound flight, squawk A2000.

When entering Bermuda, squawk A2100 until ATC assigns you another transponder code. [†]

6.	All aircraft that have been cleared on organized tracts are exempt from transmitting meteorological observations.  The reason for this is that meteorological services do routinely collect such information from other equally reliable sources.

7.	Whenever a change of flight level or Mach number will improve flight efficiency, request it from ATC.

Let us now simulate a position report at a hypothetical oceanic position [*]:

**N163H:**  Santa Maria, Santa Maria, NO VEM BER WUN SIX TREE HOH TEL.

**Santa Maria:**  NO VEM BER WUN SIX TREE HOH TEL, Santa Maria, go ahead.

**N163H:**  Santa Maria, Santa Maria, NO VEM BER WUN SIX TREE HOH TEL, position FOW-er, FIFE, WUN, WUN North (45:11N) TREE, FIFE, WUN, FIFE West (35:15W), ZEE RO AIT WUN FIFE (08:15), flight level TREE FIFE ZEE RO (FL 350), FIFE, TOO, WUN, FIFE North (52:15), TREE, SEV-en ZEE RO, ZEE RO West (37:00W), ZEE RO NIN-er WUN TOO (09:12), SIX, ZEE RO, TOO, FIFE North (60:25N), TREE, NIN-er, WUN, FIFE West (39:15W), ZEE RO NIN-er FIFE AIT (09:58), Mach DAYSEEMAL AIT TOO (M .82) (over).

**Santa Maria:**  Roger, NO VEM BER WUN SIX TREE HOH TEL, Santa Maria, estimates copied, maintain flight level TREE FIFE ZEE RO (FL 350).  Call again FIFE, TOO,

---

[†] Contact the current ACs and Notams for current Transponder codes to use.

[*] Applies to the North Atlantic, see note 10 for Pacific Ocean Crossings.

32

WUN, FIFE North  (52:15N), TREE SEV-en ZEE RO ZEE RO West (37:00W) over.

**N163H:**  Roger, Santa Maria, NO VEM BER WUN SIX TREE HOH TEL.


## Distress and Related Emergency Messages[11]


      We shall now introduce phraseology needed to cope with emergency or distress, urgency, and safety.  An aircraft is in a distressing situation if it is threatened by grave and imminent danger and requests immediate assistance.  The distress signal issued by the aircraft is "Mayday".  In a distressing situation the aircraft must transmit a distress message consisting of the following:


1.        The word "Mayday" repeated thrice, followed by
2.        The phrase, "This is" which in turn is followed by the call-sign of the aircraft repeated thrice.
3.        The main body of the message, given in the following order:
4.        Nature of the distress
5.        Pilot intention
6.        Position
7.        Heading, stating whether "True" or "Magnetic" is being maintained
8.        True air speed (TAS)
9.        Type of assistance needed and
10.        Any additional information
11.        The phrase, "This is" followed by the call-sign of the aircraft, and the word "over".

For example, let us treat our simulated flight as if we had an emergency:

**N163H:**  Mayday, Mayday, Mayday.  This is NO VEM BER WUN SIX TREE HOH TEL, NO VEM BER WUN SIX TREE HOH TEL, NO VEM BER WUN SIX TREE HOH TEL, lost port engine, starboard engine erratic, losing altitude, will attempt a forced landing.  Position, two miles west of PAH PAH HOH TEL beacon, heading two FOW-er ZEE RO (240°M) degrees magnetic, true airspeed, TREE FIFE ZEE RO (350) knots.  Request rescuc assistance. Aircraft color, red.  This is NO VEM BER WUN SIX TREE HOH TEL, over.

Note the sequence in the delivery.  All distress messages are transmitted this way.

Although you could broadcast your distress message on the existing frequency, the frequency 121.5 MHZ is the designated emergency frequency.  If you did not change to this frequency, ATC will direct you to change to it.  Another frequency for distress is 243 MHZ for aircraft also equipped to transmit on 121.5 MHZ.

## Urgency Messages

The urgency signal is "Pan".  An urgency situation arises when a very urgent message is required to be transmitted regarding the safety of another aircraft, ship, other vehicle, or even a person.  The following is what an aircraft hearing your distress message will transmit to Iko Approach in case you lost VHF line of sight as you lost altitude, and were not heard by Iko Approach.  Assume the aircraft call-sign is **N177V.**

**N177V:**  Pan, Pan, Pan.  This is NO VEM BER WUN SEV-en, SEV-en VIKTAH.  NO VEM BER WUN SIX TREE HOH TEL lost port engine, starboard engine erratic, losing altitude, will attempt a forced landing.  Position, two miles west of PAH PAH HOH TEL beacon, heading two FOW-er ZEE RO degrees magnetic, true airspeed, TREE FIFE ZEE RO knots, request rescue assistance.  Aircraft color, red. This is NO VEM BER WUN SEV-en, SEV-en VIK TAH, over.

You will have noticed that this is a verbatim reproduction of the distress message communicated by N163H except for the

34

use of "Pan" in place of "Mayday". In response to this transmission, Iko Approach will address N163H.

**Iko Approach:**  Roger, NO VEM BER WUN SIX TREE HOH TEL, Iko Approach copied. NO VEM BER WUN SIX TREE HOH TEL lost port engine, starboard engine erratic, losing altitude, attempting a forced landing.  Heading two FOW-er ZEE RO degrees magnetic, true airspeed TREE FIFE ZEE RO knots, two miles west of PAH PAH HOH TEL beacon, aircraft color, red, alerting Search and Rescue, over.


## Safety Messages


Safety messages concern the safety of air navigation or any important meteorological phenomenon.  The safety warning comes in the form of a signal "Security".  Like the urgency message, it is transmitted three times followed by the relevant safety warning.  It has priority over normal messages.  Put differently, it is third in priority after Distress and Urgency messages.


For our purposes, we shall now assume that you regained power on both your engines and are now ready to climb back to flight level two SEV-en ZEE RO (FL270) to resume normal navigation.  You probably got your engines iced up so you lost one while the other was erratic from ice ingestion.  As you lost altitude, you descended into warmer air and you were able to recover smooth operation on both engines.  You now choose a lower altitude for fear of ice.  We shall assume that the other airplane is still in the vicinity, to introduce the procedure for relaying a message from one station to another.

### Relaying a message on R/T

As you already are aware, you (N163H) descended too low for VHF contact with Iko Approach and having recovered your engines, you are now ready to continue your journey. So, you call the aircraft that earlier relayed your distress message to relay for you again.

**N163H:** NO VEM BER WUN SEV-en, SEV-en VIKTAH, NO VEM BER WUN SIX TREE HOH TEL, request relay to Iko Approach, over.

**N177V:** NO VEM BER WUN SIX TREE HOH TEL, this is NO VEM BER WUN SEV-en SEV-en VIKTAH, go ahead for relay.

**N163H:** Roger, NO VEM BER WUN SIX TREE HOH TEL have re-ignited port engine, starboard engine now normal, resuming normal navigation, requesting flight level two SEV-en ZEE RO (FL270) due icing at higher levels, maintaining flight level SEV-en ZEE RO (FL70)[*] (over).

**N177V:** Roger, NO VEM BER WUN SIX TREE HOH TEL, copied, *break,* Iko Approach, NO VEM BER WUN SEV-en, SEV-en VIKTAH, relaying for NO VEM BER WUN SIX TREE HOH TEL, re-ignited port engine, starboard engine now normal, resuming normal navigation, requesting flight level TOO SEV-en ZEE RO due icing at higher levels, now maintaining flight level SEV-en ZEE RO (over).

Note the use of the word "break" in the preceding relayed message to break or discontinue the communication with N163H in order to start one with Iko Approach. That is the major addition to a relayed message, besides the use of the word "relaying". Gain more familiarity with the use of 'break" in the following transmission as we continue our simulation.

---

[*] The airspace structure in the US makes it impossible to have flight levels below 18,000 feet. This is not the case in other countries.

**Iko Approach:** Roger, NO VEM BER WUN SEV-en, SEV-en, VIKTAH, Iko Approach, copied, *break*, NO VEM BER WUN SIX TREE HOH TEL re-cleared flight level TOO FIFE ZEE RO (FL250) to IN DEE AH KEY LOH (IK) VOR, no delay, runway two FOW-er, Quebec NO VEM BER Hotel, WUN ZEE RO WUN, TOO Hectopascals, ATC time, WUN TREE . Flight level TOO SEV-en ZEE RO not available due traffic over.

Note how the approach acknowledges the message from N177V but then breaks away to transmit a clearance to N163H. N177V will now pick up the clearance and relay to N163H. Notice that the clearance includes arrival terminal information because apparently this airport does not have ATIS arrival.

**N177V:** Iko Approach, NO VEM BER WUN SEV-en, SEV-en VIKTAH copied, *break*, Iko Approach clears NO VEM BER WUN SIX TREE HOH TEL flight level TOO FIFE ZEE RO to IN DEE AH KEY LOH VOR, no delay, runway TOO FOW-er, Quebec NO VEM BER Hotel, WUN ZEE RO WUN TOO, ATC time, WUN TREE , flight level TOO SEV-en ZEE RO not available due traffic, over.

**N163H:** Roger, NO VEM BER WUN SEV-en, SEV-en VIKTAH, NO VEM BER WUN SIX TREE HOH TEL copied, cleared flight level TOO FIFE ZEE RO to IN DEE AH KEY LOH VOR, no delay, runway TOO FOW-er, Quebec NO VEM BER Hotel, WUN ZEE RO WUN TOO, ATC time checked. Flight level TOO SEV-en ZEE RO not available due traffic.

**N177V:** NO VEM BER WUN SIX TREE HOH TEL, that is correct, break, Iko Approach, your clearance relayed, NO VEM BER WUN SEV-en, SEV-en VIKTAH.

**Iko Approach:** Roger, NO VEM BER SEV-en VIKTAH, Iko Approach, NO VEM BER TREE HOH TEL to contact Iko Approach when within range (over).

**N177V:** Iko Approach, roger, *break*, NO VEM BER TREE HOH TEL to contact Iko Approach when within range (over).

**N163H:**  Roger, contact Iko Approach when within VHF range, NO VEM BER TREE HOH TEL.

You will notice that towards the end of this transmission we were using abbreviated call-signs.  When this is done, it has to be initiated by the ground station and there is a system to it.  You use the first and last two characters or only the last two characters when the first character, which represents the nationality of the aircraft is not in doubt.  Abbreviated call signs should never be used at an initial call or at any time when other aircraft call signs have similar numbers/sounds or identical letters/numbers.

A word about aircraft registration marks[12] and call-signs.  They consist of five character displays.  There are two parts to it, the first part which may consist of one or two characters represents the nationality while the second part which is usually separated from the first by a hyphen, is the registration mark.  This latter part could contain three or four characters.  Nations with one letter standing for their nationality, e.g., France, Great Britain, United States, and others, have four character registration marks while those whose nationality is represented by two characters have each a three-letter registration mark.  This ensures a total of five characters for the nationality and registration mark.

**Examples:**

| Country | Nationality Marks | Registration Marks |
|---|---|---|
| Ghana RFC | 9G- | Any three letters, e.g., |
| Nigeria RSC | 5N- | Any three letters, e.g., |
| France VRSC | F- | Any four letters, e.g., |
| GT. Britain VRSC | G- | Any four letters, e.g., |
| USA | N- | Any alphanumeric combination of 4 digits and letters, e.g., N-177V |

Full nationality and registration marks would look like the following for the above examples, respectively:

9G-RFC, 5N-RSC, F-VRSC, G-VRSC, and N-177V.

Aircraft are also designated by their International Air Transport Association (IATA) call-signs. These are also reducible to five characters for instance Nigeria Airways flights are designated as WT- followed by three character flight numbers. You can have WT-863 for instance. So also can you have KL-863 for KLM Royal Dutch Airlines flight. Both can be abbreviated using the first two characters, that is, WT or KL followed by the last two letters, or if both flights are not approaching the same airport at the same time, the last two digits only may be used.
In addition to all these, aircraft are also called by their airline call-signs like "Skypower" for Nigeria Airways, "Clipper" for Pan American Airways, or "Speedbird" for British Airways, followed by their flight numbers. Never mind two of these airlines are now history.

## Frequency for Relayed Messages

Some of the transmissions made prior to this point have been done on the emergency frequency because the aircraft was in distress. However, under normal circumstances when a relay is required, another frequency aside from the current ATC frequency is chosen so as to avoid congestion on the existing frequency. This frequency could be the company operations frequency of either airliner, or on 126.9 MHZ. The latter frequency is used for blind transmissions in places where traffic is sparse and navigational aids are far apart, such as across deserts and seas. This brings us to broadcasts made by aircraft.

## Blind Transmissions and Broadcasts

It is at times expedient to make blind transmissions, that is, transmissions you make even though you are not sure you are being heard by the station you are targeting simply because the station does not respond. The problem could be that you have a transmitter failure or the station has a receiver failure. If other stations are receiving your transmission, then it is likely that it is the ATC station that has a problem with reception or transmission. If the station does respond to other stations, then it may be your equipment that is faulty. If on the other hand other stations tuned to the same frequency can receive you and you can communicate with them but you still cannot receive the station to which you are transmitting, then the problem is the station's. In this case, you are not alone in not being able to contact the station. On such occasions, you should transmit blind. If it is a failure on your part, there are procedures that you would have to adopt with regard to flying the aircraft. Those are outside the scope of this book but can be found in the radio failure procedures section of the Air Law and Regulations of the country concerned. However, the mechanics of blind transmission are what we would discuss here. In blind transmissions, your goal is to be heard so that other aircraft flying in your neighborhood would maintain separation from your traffic or ground stations responsible for your area of navigation would provide separation from other traffic. There is no better way to be heard than to repeat yourself since you cannot shout for better reception. You will have to state that you are transmitting blind.

In broadcasts, you transmit just as in blind transmissions, but you will state that you intend your message to be consumed by all aircraft. In the examples that follow, we shall assume that an aircraft is flying across the Sahara desert and is transmitting blind followed by a broadcast of the same message to all stations. The reason for

40

a blind transmission may not always be radio failure but may also be due to loss of radio contact due to range.

**KL-867:** Kano center, KLM AIT SIX SEV-en, transmitting blind, transmitting blind, checked Kano FIR, WUN fife, Kano FIR, WUN fife, flight level TREE SEV-en ZEE RO (FL 370), flight level TREE SEV-en ZEE RO (FL 370), estimating abeam Ndjamena TOO AIT (:28), estimating abeam Ndjamena TOO AIT (:28), Abeche next, Abeche next, wind, WUN WUN ZEE RO, diagonal SIX FIFE (W/V 110/65), wind, WUN WUN ZEE RO, diagonal SIX FIFE (W/V 110/65), temperature, minus TREE  SIX (-36°C), temperature minus TREE  SIX (-36°C), over.

In the above transmission notice that the full designation—KLM rather than KL is used.  Note also that the ordinary letters are used rather than the alphabet word designation, e.g., "Kay, El, Em" respectively rather than Kilo, Lima, Mike.  This is done when the letters are also ordinarily used as referents for the same thing.  The above is the blind transmission. The broadcast now follows.

**KL-867:**  All stations, all stations, KLM AIT SIX SEV-en, KLM AIT SIX SEV-en, east bound, from Niamey to Jeddah, east bound from Niamey to Jeddah, checked Kano FIR WUN FIFE (:15), Kano FIR WUN FIFE (:15), flight level TREE SEV-en ZEE RO (FL 370), flight level TREE SEV-en ZEE RO (FL 370), estimating abeam Ndjamena, TOO AIT (:28), abeam Ndjamena, TOO AIT (:28), Abeche next, Abeche next, wind, WUN WUN ZEE RO diagonal SIX FIFE (W/V 110/65), wind, WUN WUN ZEE RO diagonal SIX FIFE (W/V 110/65), temperature minus TREE SIX, temperature, minus TREE SIX, out.

Note the use of  "over" at the end of the blind transmission in case somebody hears and is willing to relay. It is an invitation to communicate with the aircraft.  Contrast this with the use of "out" at the end of the broadcast which does not require response from any station.  The aircraft  has a message concerning her traffic for those concerned to note.

You will also notice that the direction of flight is included in the broadcast to warn other traffic that might need to know since there is no radar service for traffic separation.

Recall that you were asked to contact **Iko Approach** when within range.  Now as an exercise, write out what you expect to tell Iko Approach when you are within VHF range.  Restrict yourself to the initial call only.  You can then check your answer for correctness below.

**N163H:**  Iko Approach NO VEM BER WUN SIX TREE HOH TEL, Iko approach, maintaining flight level TOO FIFE ZEE RO (FL 250), revised estimate PAH PAH HOH TEL, TOO WUN(:21), Iko TREE ZEE RO (:30) (over).

**Iko Approach:**  NO VEM BER WUN SIX TREE HOH TEL, Iko Approach, copied, call ready for descent (over).

It is obvious that the first thing that the approach would like to hear from N163H after the establishment of communication contact is the revised estimate for her station, since the original estimate had gone awry owing to the engine failure.  It is therefore good airmanship to come up with this information as soon as communication contact is established.  It is the more important where there is no radar service.  Remember that Iko Approach did not want to over burden N177V with the additional work of relaying for you hence she did not ask her to relay your estimates to her.  She would have asked her to relay.

**N163H:**  Iko Approach, NO VEM BER WUN SIX TREE HOH TEL, request descent (clearance).

Clearance is in parenthesis since it is not really necessary to say it

**Iko Approach:**  Iko approach clears NO VEM BER WUN SIX TREE HOH TEL to descend to flight level WUN ZEE RO ZEE RO (FL 100), rate of descent, TREE TOUSAND feet per minute (3,000 ft/min) or more.  Call approaching flight level WUN ZEE RO ZEE RO (FL 100) for further descent.

**N163H:**  Roger, cleared flight level WUN ZEE RO ZEE RO (100), rate of descent TREE TOUSAND feet per minute or more, call approaching, NO VEM BER TREE HOH TEL..

**Iko Approach:**  NO VEM BER TREE HOH TEL, that is correct.
With such a high rate of descent, if you wait until you are about a thousand feet to go before requesting further descent, you are likely to burn a lot of gas waiting for clearance when proper planning should have given you an uninterrupted descent.  You should accordingly plan to request further descent when you are just more than the requested rate of descent to your "cleared to" altitude.  In this example for instance, you would request further descent when you are at about flight level 140 since you are descending at the rate of 3000 feet per minute and flight level 140 is four thousand feet to go to flight level 100.

**N163H:**  Iko Approach, NO VEM BER TREE HOH TEL, descending out of flight level WUN FOW-er ZEE RO (FL 140) for WUN ZEE RO ZEE RO (100), request further descent (over).

**Iko Approach:**  Roger, NO VEM BER TREE HOH TEL, you are re-cleared IN DEE AH KEY LOH VOR, TREE FIFE ZEE RO ZEE RO (3,500) feet, call out of flight level WUN ZEE RO ZEE RO (FL 100) (over).

**N163H:** Roger, NO VEM BER TREE HOH TEL, re-cleared TREE FIFE ZEE RO ZEE RO (3,500) feet, call out of flight level WUN ZEE RO ZEE RO (FL 100).

**Iko Approach:** NO VEM BER TREE HOH TEL, Iko Approach, that is correct.

**N163H:** Iko Approach, NO VEM BER TREE HOH TEL, out of flight level WUN ZEE RO ZEE RO (FL 100) for TREE FIFE ZEE RO ZEE RO (3,500) feet (over).

**Iko Approach:** NO VEM BER TREE HOH TEL, contact Iko Radar now on WUN TOO TOO dayseemal TREE (122.3 MHZ) (over).

**N163H:** Roger, WUN TOO TOO dayseemal TREE. NO VEM BER TREE HOH TEL.

**N163H:** Iko Radar, NO VEM BER WUN SIX TREE HOH TEL, descending out of flight level WUN ZEE RO, ZEE RO (FL 100), approaching TREE FIFE ZEE RO, ZEE RO (3,500) feet, over.

You can see that in your first contact with Iko Radar, you did not start by first calling radar and then calling and giving your position report. Rather, you went on straight away to tell radar what you were doing. This procedure helps expedite action since radar would immediately have you on her scope as soon as you are handed over. It helps when you tell the controller what altitude you are passing or what maneuver you are executing.

**Iko Radar:** NO VEM BER WUN SIX TREE HOH TEL, squawk alpha TREE WUN TOO TOO (A3122) with Charlie (over).

**N163H:** Roger, NO VEM BER WUN SIX TREE HOH TEL, alpha TREE WUN TOO TOO with Charlie.

**Iko Radar:** NO VEM BER TREE HOH TEL, radar contact, cleared to descend to TOO TOUSAND (2,000 ft.) feet on QNH (altimeter setting)[*] WUN ZEE RO WUN TOO (1012), and on reaching IN DEE AH KEY LOH, cleared for a VOR approach, runway TOO FOW-er, obstacle clearance limit, TREE AIT FIFE (385) feet on QNH WUN ZEE RO WUN TOO (1012), read back, over.

**N163H:** Roger, NO VEM BER TREE HOH TEL, cleared TOO TOUSAND feet on WUN ZEE RO WUN TOO, and on reaching IN DEE AH KEY LOH, cleared for a VOR approach runway TOO FOW-er, obstacle clearance limit (OCL) TREE AIT FIFE feet on QNH (altimeter setting) WUN ZEE RO WUN TOO.

In the response, the first QNH is missing. This was intended since it helps to shorten the message. Notice that the letters "Q, N, H" are used rather than their word designations, KEH BECK, NO VEM BER, and HOH TEL respectively. This is the practice when there is no ambiguity and when it will save time to use the letters rather than their International Civil Aviation Organization (ICAO) designations. In a similar vein, we used the letters "K, L, M" rather than KEY LOH, LEE MAH, and MIKE respectively when referring to the KL-863 flight. Note that in the QNH given, it is not stated that it is millibars or hectopascals. This omission helps save time since it is clear that "millibars" or "Hectopascals" were intended. Inches of mercury may not be this high. Remember that it is the intention in this book to reproduce as much of the real life situations as possible. You want to know what to expect to hear and what to say in real life and that is exactly what this book is about.

---

[*] In the US the phrase "altimeter setting" is used rather than "on QNH" preferred in Europe and Africa.

In instrument "let downs", you will need to report when over the facility outbound for the procedure turn. You will also report the completion of procedure turn inbound and attainment of minimums. In addition, you will report "runway in sight" or "runway not in sight, commencing a missed approach" whichever is applicable.

**N163H:** Iko radar, NO VEM BER TREE HOH TEL, IN DEE AH KEY LOH, TREE  ZEE RO (:30), TOO TOUSAND (2,000) feet, beacon outbound, call you procedure turn complete inbound, over.

Notice the sequence in the preceding transmission—first, the station being addressed, then your call-sign, next, your position, your altitude and time, and finally, what you are doing. This is the standard approach to reporting positions. It is no different when shooting an approach.

**Iko Radar:** Roger, NO VEM BER TREE HOH TEL, continue your approach.

**N163H:** Iko Radar, NO VEM BER TREE HOH TEL, procedure turn complete inbound, over.

**Iko Radar:** NO VEM BER TREE HOH TEL, roger, obstacle clearance limit, TREE  AIT FIFE (385) feet altimeter setting WUN ZEE RO WUN
TOO (1012) millibars, report runway in sight <u>to tower</u> on WUN WUN AIT dayseemal WUN (118.1 MHZ) (over).

**N163H:** Roger, NO VEM BER TREE HOH TEL, WUN WUN AIT dayseemal WUN (118.1 MHZ) (out).

If you see the runway at your minimums, then you continue your approach to the runway reporting the fact to the tower who then clears you to land. However, if you do not see the runway, you make a missed approach and continue on radar frequency without switching to tower. As long as you are at or above minimums, you are a radar/approach traffic.

**N163H:** Iko Tower, NO VEM BER WUN SIX TREE HOH TEL, WUN WUN AIT dayseemal WUN (over).

Notice the use of the full call-sign of the aircraft in the initial contact with the tower. Always remember to use your full call-sign on initial contact.

**Iko Tower:** NO VEM BER WUN SIX TREE HOH TEL, Iko Tower, go ahead.

**N163H:** Iko Tower, NO VEM BER TREE HOH TEL runway TOO FOW-er (R/W 24) in sight (over).

**Iko Tower:** NO VEM BER TREE HOH TEL, roger, I have you in sight, (check gears down and locked), cleared to land runway TOO FOW-er (R/W 24). Surface wind, TOO TOO ZEE RO diagonal WUN FIFE (W/V 220/15) knots (over).

**N163H:** Cleared to land, runway TOO FOW-er (R/W 24), NO VEM BER TREE HOH TEL.

Notice that the omission of the word "degrees" in the surface wind given helps reduce congestion on the frequency since it is understood that the first part in the surface wind advisory contains the wind direction. The word "knots" in the second part may be omitted without compromising the meaning.

In the above simulation, notice the words in parenthesis. The tower controller may or may not include them in her transmission; hence they are in parenthesis. However, if you are not expecting to hear such words they may take you by surprise and so you may not understand what was said. They are included as a guide to what may be said. If there is any cautionary advice, for instance concerning runway conditions, or wind conditions on the final approach path that may be hazardous, it is likely to be given here too.

It is always good airmanship to read back the runway to which you have been cleared to approach or land as was done above. The procedure ascertains that you are landing on the correct runway, especially in places where there are multiple runways. In such places it is quite easy to acknowledge clearance to land and then land on the wrong runway. Pilots have been known to land on runways under construction as a result of the proximity of the landing runway to the one under construction. If you are aligned with the wrong runway, when you read back the clearance the ATC may realize the error and ask you to make a missed approach and give you a re-clearance. If you did not read back the runway in your read back, then the ATC may not realize that you are incorrectly aligned with the runway to which you were cleared.

**Iko Tower:** NO VEM BER TREE HOH TEL, touch down TREE FIFE, next convenient right/left. Contact ground on WUN TOO WUN dayseemal NIN-er (121.9 MHZ) when cleared off the runway (over).

Ground control frequency is used to control ground traffic at airports where the volume of traffic is too much for the tower to handle.
If you are carrying out a missed approach, radar/approach will give you clearance to join the holding pattern or proceed for another approach depending on traffic or the laid down procedure, unless of course, you elect to divert.

**N163H:** Iko Radar, NO VEM BER TREE HOH TEL, minimums, runway not in sight, missed approach (over).

**Iko Radar:** Roger, NO VEM BER TREE HOH TEL, call again IN DEE AH KEY LOH VOR joining the hold (over).

The controller may in the alternative request to know your intentions after you report that you are carrying out a missed approach before giving you missed approach instructions.

**Iko Radar:** Roger, NO VEM BER TREE HOH TEL on missed approach, request your intention after missed approach (over).

**N163H:** Iko Radar, NO VEM BER TREE HOH TEL, request ILS approach, runway TOO FOW-er (over).

In the above case, you chose to shoot an ILS approach rather than repeat the VOR approach or divert. Note here also that abbreviations like ILS, VOR, QNH, etc., that stand for well known facilities, settings, etc, do not have to be read out in their ICAO letter designations since there is no ambiguity when they are called by their standard names. For instance we do not say VIK TAH OSS CAH ROH ME OH for VOR, but simply, Vee-Oh-Ar.

**Iko Radar:** Roger, NO VEM BER TREE HOH TEL, you are cleared ILS approach, runway TOO FOW-er, no delay, call Outer Marker (OM) inbound established on the Glide Slope (G/S) (over).

You now continue your approach to the Outer Marker (OM) and report, at which stage the radar controller will give you the Decision Height (DH) with full ILS, thus:
**N163H:** Iko Radar, NO VEM BER THREE HOH TEL, Outer Marker inbound, established on the Glide Slope (over).
**Iko Radar:** Roger, NO VEM BER TREE HOH TEL Decision Height (DH) TOO ZEE RO SIX (206) feet, QNH WUN ZEE RO WUN TOO (1012) millibars. Continue your approach and report runway in sight to tower on WUN, WUN AIT dayseemal WUN (118.1 KHZ) (over).

In a Category II approach, you will be asked to report at the Middle Marker before reporting runway in sight.

**N163H:** Iko Radar, NO VEM BER THREE HOTEL, Middle Marker inbound (over).
**Iko Radar:** NO VEM BER TREE HOTEL, Iko Radar, Decision Height 206 feet, QNH WUN ZEE RO WUN TOO (1012) millibars. Continue your approach and report runway in sight to tower on WUN, WUN AIT dayseemal WUN (118.1 KHZ) (over).

## Glossary

**Affirmative:** This word stands for "yes". To a question requiring the word "yes", you should use "affirmative" instead. You may also use "affirmative" when your intention is to correct. In this regard, you put the word "affirmative" before the correct version of your incorrect message, e.g., The controller has erroneously called you NO VEM BER WUN SIX TOO HOH TEL but your correct call-sign is NO VEM BER WUN SIX TREE HOH TEL. To correct this you will say: "My call-sign negative NO VEM BER WUN SIX TOO HOH TEL, **affirmative** NO VEM BER WUN SIX TREE HOH TEL.

**Break:** This word is used to indicate a break in transmission. If you were talking to a facility and you wanted to interrupt it in order to transmit to some other station, use the word "break", then start transmitting to the second party. Good examples are in the section on relayed messages.

**Charlie:** This is often used in some French speaking countries to mean affirmative, that is, correct or "yes". A less restricted use for it is with reference to transponder where it stands for a transponder with altitude reporting capability. A favorite question using Charlie, is "Do you have Charlie?" This simply means "Do you have a transponder with altitude reporting capability?"

**Confirm:** When in doubt, you should use this word to clear such doubts by asking the other party to **confirm** his or her message. For instance, in the example under "correction" below, having corrected your estimate for Tango Juliet (TJ), the controller may ask you to "confirm your estimate for TANG GO JEW LEE ETT ZEE RO AIT FIFE, FIFE, negative ZEE RO AIT FOW-er SEV-en".

**Correct:** This is an abbreviation of "That is correct" which means that your repetition or read back of my message was correct. It is often used by ATC to affirm the correctness of clearances read back to them. It could, of course, be used in other situations.

**Correction:** When you have made a mistake in a message you sent out, this word enables you to correct that message. You do not say "I beg your pardon" or use other apologetic words, all you say is "correction" followed by the correct version of your earlier message. For example, you passed an estimate of 08:47 for position Tango Juliet when in fact you wanted to pass an estimate of 08:55. You should now say, "Correction, estimate TANG GO JEW LEE ETT, ZEE RO AIT FIFE, FIFE, negative ZEE RO AIT FOW-er SEV-en".

**Hold Position:** This means that you should remain in your present position, that is, remain where you are now. It may be given by ATC to direct you not to take off yet, or it may be given at any time that your further movement on the maneuvering area may be hazardous.

**Hold Short:** This is a command also used by ATC to instruct aircraft on the tarmac to keep off certain areas, like the active runway or an active taxi way.

**Negative:** This is the opposite of affirmative; It is used in the opposite sense too. You will already have noticed its use under "correction" above.

**Out:** This is used to terminate a conversation when a reply is no longer expected from the other party. For example, see the response in **readability check** on page 3.

**Over:** This terminates a message that requires a response from the other party. It indicates that a response is now expected from the other party. It is a cue to the other party to start responding to the dialogue. Ideally, it should end every conversation except those that end with "out", however, this is not done in order not to congest the frequency with too many understood "over".

**QDM:** This is a Q-code, a bunch of British codes which are gradually dying out. It is a magnetic heading to steer to a station in no wind condition.

**QDR:** This is your magnetic radial when flying away from a VOR station.

**QNH:** This is the barometric pressure you set on your altimeter which gives you the airport elevation when you are on the ground. E.g., if you set a QNH of 1011 Millibars in the pressure window of your altimeter, when you are on ground, the altimeter will register the airport elevation above mean sea level.

**Readback:** This is an instruction to repeat what has been said. It is often used by ATC to ascertain the correct reception of a message transmitted to an airplane. It can also be used as a noun to refer to what has been so repeated.

**Request:** This word is very handy. It is like a master word. It is a shorthand for "please, I wish to …" or "could you please permit me to …", etc. You will find that in R/T we do not use the word "please" since it is implicit in "request".

**Roger:** This word means your message is understood, and I will comply. It is a very commonly used word for responding to messages during conversations between two parties.

**Squawk:** This is a directive to select a given code on your transponder. For instance, "Squawk alpha TREE  ZEE RO TREE , TREE " means that you should select mode Alpha 3033 on your transponder and switch on the equipment. For a message which instructs you to "squawk ident", you would press the "ident" button on the transponder momentarily.

**Standby:** This means wait. The number of minutes to wait is also usually designated for up to five minutes. The numbers one to five are used to designate the waiting period in minutes estimated by the speaker, with one being equal to one minute wait and five, being equal to a wait of five minutes. A statement combining the word standby and any of the five digits is an instruction on the number of minutes of waiting to expect. Thus we have standby one, standby two, etc, meaning to wait for one minute, wait for two minutes, etc, respectively. Standby one would be equivalent to saying, "hold on a moment".

**Vectors:** These are magnetic headings to steer given to air traffic at a radar controlled airport. At times they are given only on request to aid the pilot in positioning for a landing. It may also be given after take-off to help in traffic separation.

**Wilco:** This conveys the same message as "roger" noted earlier. It is a shortened form of "will comply". It is a common response to ATC instructions.

**Words Twice:** This is an instruction to say each word in a message twice for clarity. It is used if the recipient has difficulty in reception.

## **<u>Some Notes on I/R Renewal Tests</u>**

The instrument rating (I/R) renewal test is given every thirteen months to keep an I/R in vigor in countries like the United Kingdom and other Commonwealth countries. In the United States, the I/R test is given only as an initial rating and no renewal is necessary as long as the pilot flies a specified minimum number of hours annually to keep current. If at any time the minimum flight hours have not been flown, the erring pilot will undergo a full instrument rating check-out including ground school and a flight test before being allowed to exercise the privileges of an instrument rated pilot again. Thus the I/R renewal test procedure outlined in this book will focus on the requirements of a Commonwealth licensed pilot.

It is necessary to have an audit of all the requirements that the examiner would want you to have prior to you taking the exam. Thus you should be sure to have the following when you turn up for the test:
1.      A valid license that has been signed.
2.      Valid medical certificate with any restrictions met, for example the stipulation to have an extra pair of glasses if you use one.
3.      Necessary aircraft types in Part 1 or 2.
4.      R/T license valid and signed.
5.      Instrument Rating Certificate valid.

There are three parts to the normal instrument rating renewal test:
a.      Section 1 and Sub-section A: Pre-flight, Take-off and Climb
b.      Section 2: Airways Procedure
c.      Section 3: ILS and Missed Approach

The instrument rating is valid for a maximum period of 13 months and the purpose of the renewal test is to ascertain that the pilot has maintained the standard of proficiency necessary for safe operation within controlled airspace under the instrument flight rules. The maximum period within which it may be ordinarily renewed after an expiry is 5 years after which special requirements may be demanded.

## Pre-Flight

You should always assume that icing conditions prevail from ground up when you are slated for an instrument rating renewal flight test. Therefore, your pre-flight inspection should place special emphasis on demonstrating to the examiner that you are carrying out your pre-flight inspection with this assumption in mind. You should also operate the airplane as if there were icing. Be warned that any excessive time spent on a procedure that may cause the delay of other flights, could result in failure of that section of the test[13]. You will be tested in your usual pilot position during the flight test so be sure to carry out tasks appropriate to your pilot position. For instance, if you are a captain, then occupy the captain's left seat during the entire flight and if you are a first officer or co-pilot, then you will be tested with that assumption if the flight test is in a transport plane. However, if it is a light airplane, you will be tested as the pilot flying and operating from the left seat. In case of a single crew operation, you will not receive assistance from the examiner in carrying out any of the duties of the pilot. For instance, if it is a complex airplane with retractable undercarriage, you will be expected, during the flight test to select the gears up yourself.

Remember that although you may be flying with an examiner who may look older and more experienced than you are, you are the one in command of the flight once it is you who is being tested. Therefore, assume complete command of the flight in terms of the overall management of the flight. You have responsibility for initiating communication, tuning and identifying all radio aids according to company operating procedures, and for maneuvering the airplane. Of course, because you will lack visual ability to look out for collision avoidance, the examiner and any other pilots in the cockpit will take responsibility for collision avoidance.

The instrument examiner is expected to conduct the examination in a very professional manner with no undue pressure on the examinee who it is realized, is already under pressure to do well in the exam. Therefore, take your time to point out what you are inspecting and why you are carrying out the inspection.

The examiner will ask you to state the speed you will maintain for each phase and configuration of the test. Furthermore, he/she will describe the means of simulating instrument flight conditions to your understanding and if necessary, will explain how it works.

The examiner will tell you that you are to initiate all R/T calls and that tuning and identifying the navaids will be in accordance with normal company operating procedures. The examiner is expected to check your ability to tune and identify the radio aids during the ground operations. Normally you will taxi the airplane to the holding point. During taxi, you will advise the examiner of any visual checks you are carrying out. If you do not want to talk while doing the checks, you may leave talking until later when you are more comfortable.

## Take Off And Climb

Note that the examiner will remind you of the assumed cloud base and will expect you to be settled on the climb and flying by reference only to your instruments by this height. The examiner will be taking note of your airspeed and altitude control during the short period between take off and the cloud base so be sure to have full control of this sector of the flight. In fact, it is very important since it will be the first time the examiner will see you fly and as such, will be a great opportunity to create a good first impression.

## Airway Procedure

The examiner will plan the flight to assess your ability to intercept and maintain a specified inbound and outbound track by use of Automatic Direction Finder (ADF) and track inbound and outbound by Very High Frequency Omni-directional Radio Range (VOR). When you are tracking on ADF or VOR or even VDF which could also be tested, only one of the aids being tested will be available to you. If the examiner wants you to use an intersection however, a second aid would be made available. Therefore, be sure to plan accordingly. The test also includes a holding pattern. The holding pattern will be executed either enroute or just prior to the let down. You will be tested on your entry, exit, timing and tracking and correcting for wind on the inbound track. The speed at which you hold will also be of interest to the examiner. Thus you should be conversant with the various ICAO holding speeds for the different altitudes and flight levels for your type of airplane, whether turbo jet or propeller.

## ILS and Missed Approach

The examiner will observe your instrument flight on the ILS to the decision height and when you carry out an overshoot or go around on reaching the company decision height. You will be expected to continue the overshoot to the initial cleared altitude or flight level while using full R/T procedure. The examiner will point out that the ATC may alter the course of the planned I/R renewal test at any point

58

in time by issuing any additional instructions that must be complied with.

Allowance will be made for turbulence that temporarily impairs the standard of your instrument flying and the examiner will avoid testing in conditions of severe turbulence. However, in smooth air if your airspeed is more than 15 knots at any time, you could be assessed as failed. In addition if your airspeed is in error of more than 10 knots in cruise or 5 knots in climb or approach for more than 15 seconds, the examiner could consider giving you a fail point. With regard to height accuracy, if you maintain a height error of more than two hundred feet at any time, you could be given a failing point. Furthermore, if you maintain an error of 100 feet for more than 15 seconds, you will also be subject to receiving a fail point. The point in all this is that you are expected to keep trying to fly the numbers as accurately as you can without sacrificing the smoothness of the flight. The examiner will also expect you to fly within the tail wind limits of the particular airplane.

## Assessment System

There are tolerances for the instrument flight renewal test as you would imagine. However, the examiner would expect you to fly smoothly with good coordination while remaining within the tolerance limits. The following are the errors and omissions[14] that could result in a failure in applicable sections of the test.

- Failure to set the correct altimeter setting at any phase of the flight.
- Failure to check any of the flight instruments before the flight, including the compass.
- Failure to check any one of the flying, trimmer, stabilizer, or elevator controls before the flight, for range and freedom of operation in the correct sense.
- Failure to check before flight, any one of the following: pitot head(s) and heater(s); static vents all de-

icing, anti-icing, and ice warning equipment for serviceability, as much as it is possible to do so.

• Failure to use correctly any of the aforementioned items.

• Failure to obtain Air Traffic clearance whenever necessary.

• Failure to comply with ATC clearance.

• Failure to make a standard or requested position report to ATC.

• Jeopardizing aircraft control to such an extent that the examiner takes over control.

• In single crew operation, omission of any important check item prior to take off or landing.

• In multiple crew operation, failure to call for any of the appropriate check lists prior to take off or landing , or omission of any check item appropriate to the crew position of the candidate.

• Failure to identify the radio facility being used.

• Inability to settle within 5 degrees of the required tracks of the specified ADF, VDF, or VOR with a good signal at a suitable distance from the transmitter.

• Correcting track the wrong way and maintaining the error for more than 15 seconds.

• Serious ETA inaccuracy, such an ATA differing by three or more minutes from ETA on an airways leg of moderate length.

• During the instrument approach, failure to maintain appropriate safe altitude.

• During the intermediate phase of an ILS approach, inability to settle within 5 degrees of the required tracks of the published procedure between the holding facility and the decision height and whilst maintaining decision height if applicable.

• Inability to remain within half-scale deflection on the localizer during descent on the ILS from the published glidepath intersection height until overshoot begins.

- Inability to remain within half-scale deflection on the glidepath during descent on the ILS from the published glidepath intersection height until overshoot begins.
- Failure to apply, or in multiple crew aircraft to call for, the correct overshoot power during missed approach procedure.
- Failure to perform, or in multiple crew aircraft to call for, the overshoot checks during missed approach procedure.
- Failure to comply with the published missed approach procedure, as may be amended by ATC.
- Failure to check RVR against airfield minima before commencing approach to land.

During a let-down requiring an overshoot at decision height, if you fail to initiate an overshoot action within –0 ft. or +50 ft. of the decision height, you would get a failing point. On the other hand, if you carried out a let-down requiring the maintenance of a decision height, you failed to maintain the aircraft within –0 ft or +100 ft of decision height.

At the end of the flight in post-flight briefing the examiner will debrief you on your performance. The briefing would include what the major fail points are if you failed and the examiner will tell you how best to prepare for the retest. You should also note that a failure in one section of the test only requires a test in that section because the sections are individual entities. A failure in two sections requires a retest in the entire exam.

## Acknowledgments

The manuscript for this book has been on my writing desk for over 24 years. During that time, many things have happened to me. I have gone from being a bachelor and former airline training captain and authorized instrument rating examiner, to an undergraduate student, a graduate student, a professor of aeronautical science, a married man, an airline captain in a domestic carrier, a professor of psychology, a father, a truck driver, a research psychologist, a professor of psychology again, and finally a research scientist. Thus this book benefited from experiences I have had in all these callings and people in my life. One of them is my wife, Brenda Ikomi who has been very understanding and given me all the time required to deal with matters relating to the book. I also wish to acknowledge the contribution of my mentor, Dr. Robert Guion, emeritus professor of Bowling Green State University, Ohio, who read one of the early versions of the manuscript and poked some fun with my use of highfaluting words. Thus readers who find the current version of the book easier to read owe some gratitude to Dr. Guion. Other contributors include some of the instructors at the Nigerian College of Aviation Technology in the 1980s who gave me useful suggestions on the book after reading the manuscript. These included Captain Araba who after years in Nigeria Airways and having worked in other capacities for the government of Nigeria, was recently appointed the head of the Nigerian College of Aviation Technology. There was also Captain Ogunkoya and others whose names have faded from my memory after all these years. I hope they would understand and pardon me. To all who assisted with the manuscript I extend my gratitude.

In the 1980s developments in communication technology gave me the impression that radiotelephony was not going to be a thing of the future. So, I hesitated to continue with the book as there was a lot of talk about using non-voice communication to avoid some of the ghastly accidents that resulted from voice communication. When those new modes of communication did not materialize to wipe out radiotelephony, I picked up the manuscript from where I left off. One of the benefits of the long period the book spent on my desk is that I decided to add the section on instrument rating renewal tests which I would not have included but for the extra time of incubation the book had. Today, I am proud that this book I started shortly after I returned to college as an undergraduate, is finally completed.


Philip A. Ikomi, Ph.D.
Houston, TX USA

May 10, 2008

# <u>Index</u>

64

## <u>Notes</u>

[1] Phonetic alphabet, FAR/AIM Federal Aviation Administration Regulations Aeronautical Information Manual (Newcastle, WA: Aviation Supplies & Associates, Inc., 2007), p. 616.

[2] Ibid.

[3] Ibid.

[4] Ibid.

[5] Ibid.

[6] Ibid.

[7] Time, *FAR/AIM Federal Aviation Regulations Aeronautical Information Manual*. p.617.

[8] Pre-departure clearance procedures, *FAR/AIM Federal Aviation Regulations Aeronautical Information Manual* p.693.

[9] AC 91-70 Oceanic Operations, An Authoritative Guide to Oceanic Operations (09-06-94). Retrieved June 29, 2008 from http://www.faa.gov/ats/aat/ifim/ac91-70c04.htm

[10] Ibid.

[11] Distress and urgency communications, *FAR/AIM Federal Aviation Regulations Aeronautical Information Manual* p.780.

[12] Federal Aviation Regulation Part 45, Subpart C. Nationality and Registration Marks. Retrieved May 26, 2008 from: http://rgl.faa.gov/REGULATORY_AND_GUIDANCE_LIBRARY/RGFA R.NSF/0/FF3B72CF5B3DAD0C852566AB006BDE17?OpenDocument

[13] CAP 170. Notes for the guidance of authorized instrument rating examiners. London: Civil Aviation Authority (1976).

[14] Ibid.

Made in the USA
Monee, IL
07 July 2026

56551675R00044